Color & Learn: Rocks & Minerals

Table of Contents

Activity Guide 2	Sculpture 18
Rock Collecting 3	Mount Rushmore 19
Collecting Equipment 4	Volcanic Rock 20
Crystals 5	River of Rock 21
Pisolites 6	Aluminum 22
Geodes 7	Gold 23
Stalagmites and Stalactites 8	Coal 24
Banded Agate 9	Petroleum 25
Limestone 10	Water .26
Sand 11	Sorting Gemstones 27
Rock Arch 12	From Gemstone to Jewel 28
Pedestal Rock 13	Jewelry 29
Giant's Causeway 14	Jade 30
Ayers Rock 15	Jade Mask 31
Stonehenge 16	Pearls .32
Pyramids 17	

Illustrated by Elizabeth Adams

Reproducible for classroom use only.
Not for use by an entire school or school system.
EP219 • ©1999, 2003 Edupress, Inc.™ • P.O. Box 883 • Dana Point, CA 92629
www.edupressinc.com
ISBN 1-56472-219-8
Printed in USA

Activity Guide

Look for the ✿ at the top of the coloring page

Look and Find

Page 4. Find and make a X on the helmet, the spade, and the magnifying glass.

Page 10. Find and circle the pieces of shell in the piece of limestone.

Page 20. Find the obsidian and draw a red circle around it.

Page 28. Find and circle the gem as it is found in stone.

Complete the Picture

Page 13. Draw and color smaller rocks that have fallen away from the pedestal rock.

Page 14. Draw and color ocean waves splashing around the Giant's Causeway.

Page 17. Draw individual stone blocks in the pyramid.

Page 22. Draw and color objects made from aluminum that can be recycled.

Color to Match

Page 6. Color the pisolites inside the larger rock in different colors.

Page 7. Color the outside of the geode gray or brown. Color the inside the color of your favorite crystal.

Page 9. Color each band of the agate in a different color.

Page 31. Jade is colored different shades of red, gray, green, yellow, and white. Use these colors to color the jade mask.

Unscramble the Word

Page 5. Unscramble the names of three materials made of crystals.
almte   wfeskoaln   kocr

Page 11. Unscramble the names of materials found in sand.
altsab   rtuazq   lshels

Page 23. Unscramble the names of three things made from gold.
rta   onsci   yerljwe

Page 30. Unscramble the names of two kinds of jade.
adjetei   riphteen

Color & Learn Rocks & Minerals   © Edupress EP219

Rock Collecting

Rock collecting is a very popular hobby. Rock collectors find rocks in lots of unusual places. Collections are sorted based on rock type, color, texture, and weight.

Collecting Equipment

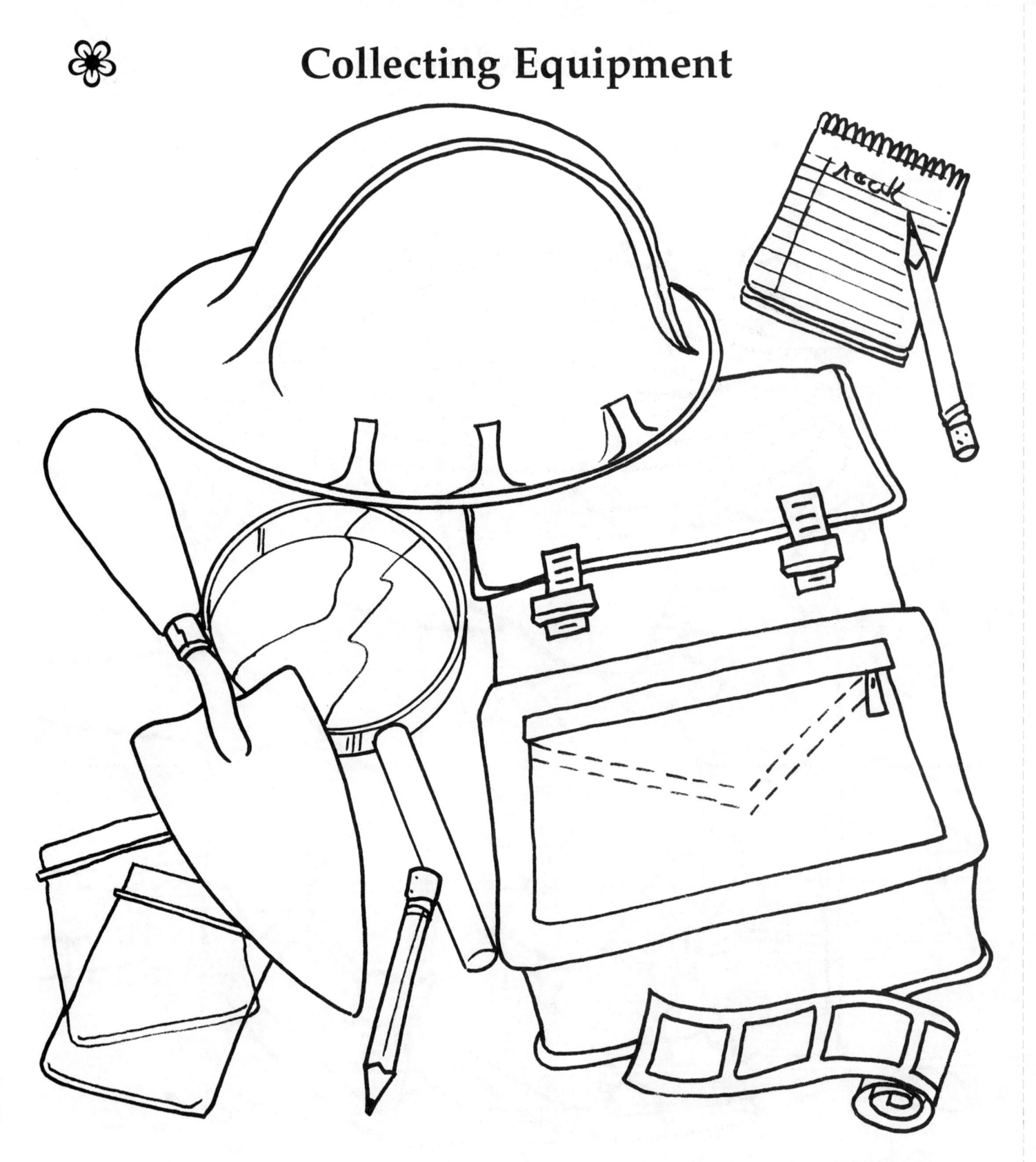

Using the right equipment and wearing the right clothes will make collecting rocks easier and safer. It is best to wear sturdy shoes and clothing. The best equipment includes a hammer, trowel, magnifying glass, eye protection, and plastic bags to store rocks.

Crystals

A crystal is a solid that is made up of atoms arranged in an orderly pattern. The arrangement of the atoms gives crystals their flat sides. Most *inorganic*, or non-living, materials are made of crystals. These include metals, rocks, snowflakes, salt, and sugar.

Pisolites

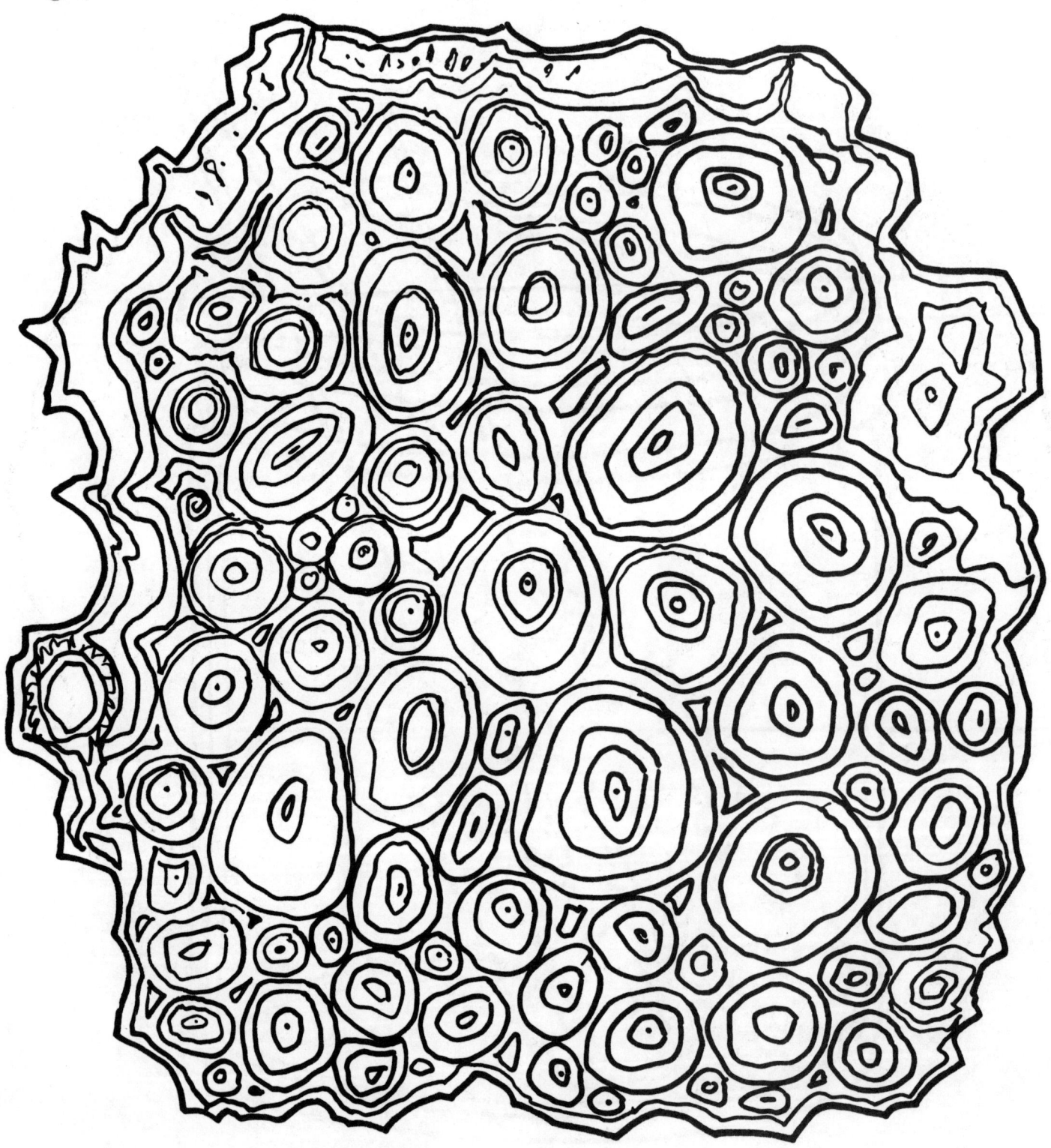

Pisolites are small ball-shaped groups of crystals. These groups are sometimes held together with another kind of rock, like limestone. The pisolites grow in layers, like the layers of an onion. When they are cut open, the different layers of crystals can be seen.

Geodes

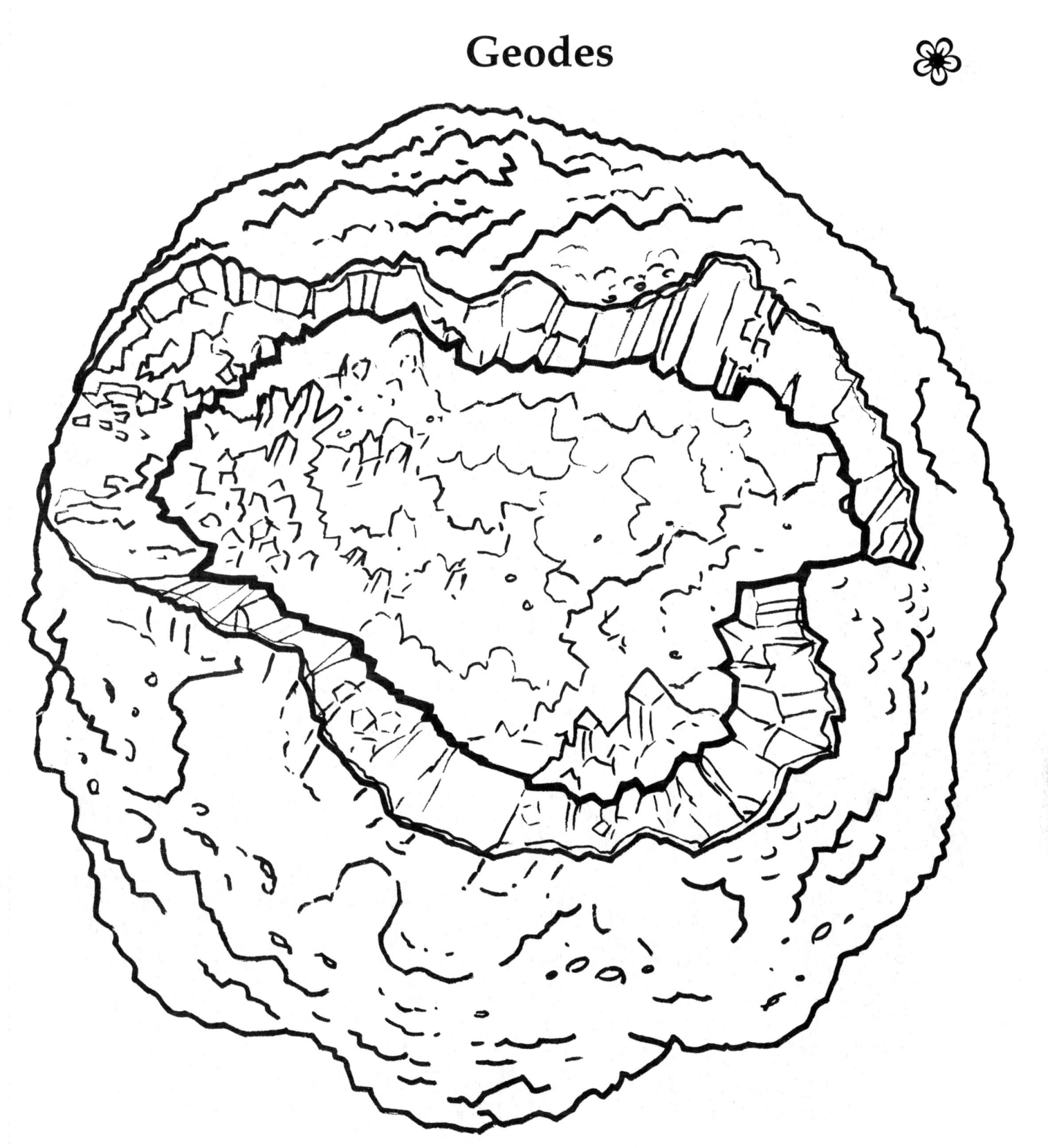

Geodes, also called thunder eggs, are unusual rocks. On the outside, the rock is usually round. When the geode is cut in half, the center is filled with crystals. Geodes form when water and silica enter a rock through cracks. The silica forms the crystals when silica becomes solid.

Stalagmites and Stalactites

The rock formations inside a cave are stalactites and stalagmites. They are formed in limestone caves, and occur when water drips through cracks, absorbing a mineral called calcite as it goes through the limestone. When the water evaporates, the calcite is left behind.

Banded Agate

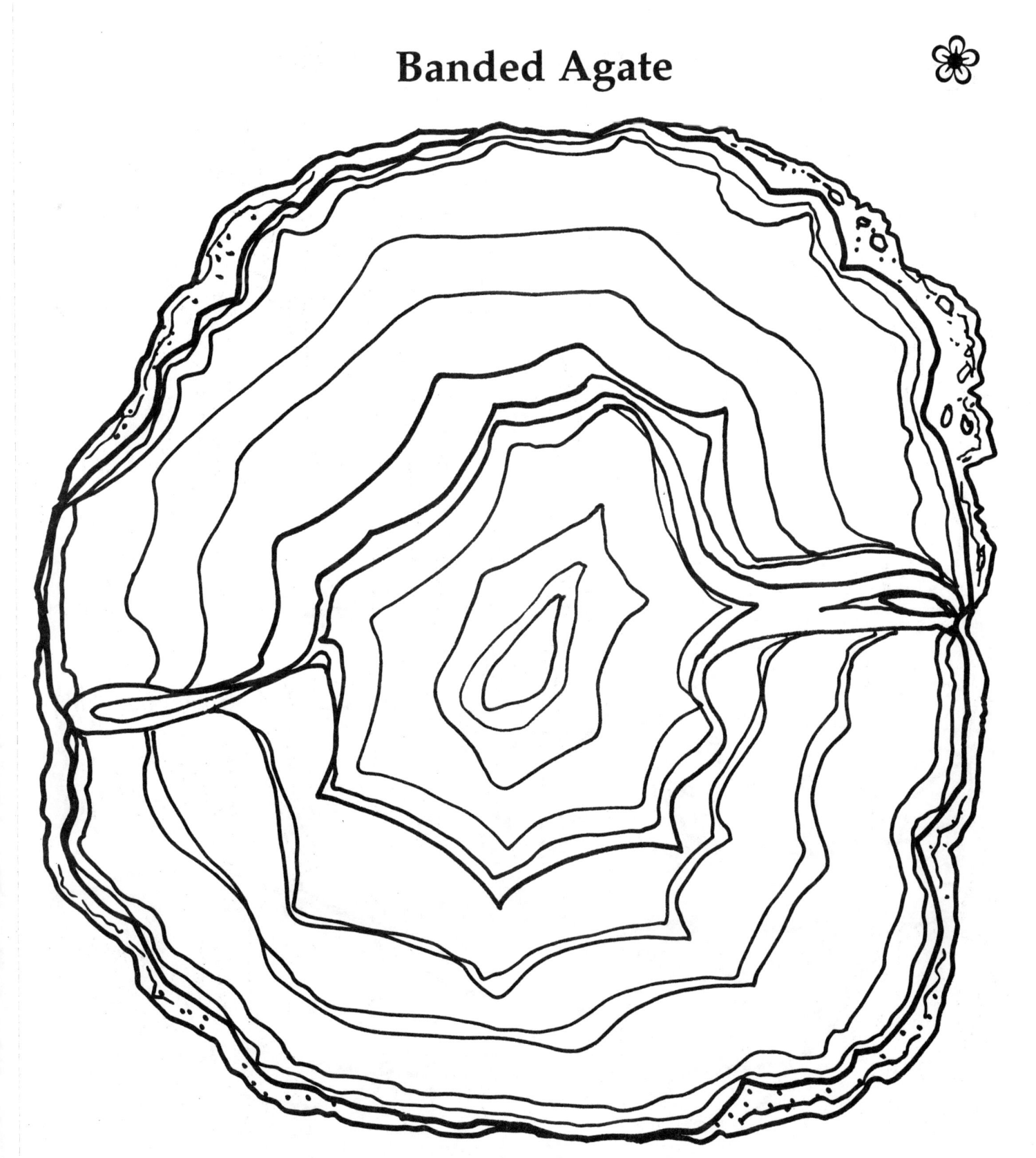

Rocks are built over time, one layer growing over the one below. Banded agate is made of layers of chalcedony, a special kind of quartz that does not show individual crystals. The many colors that are found in banded agate make it a very beautiful stone.

Limestone

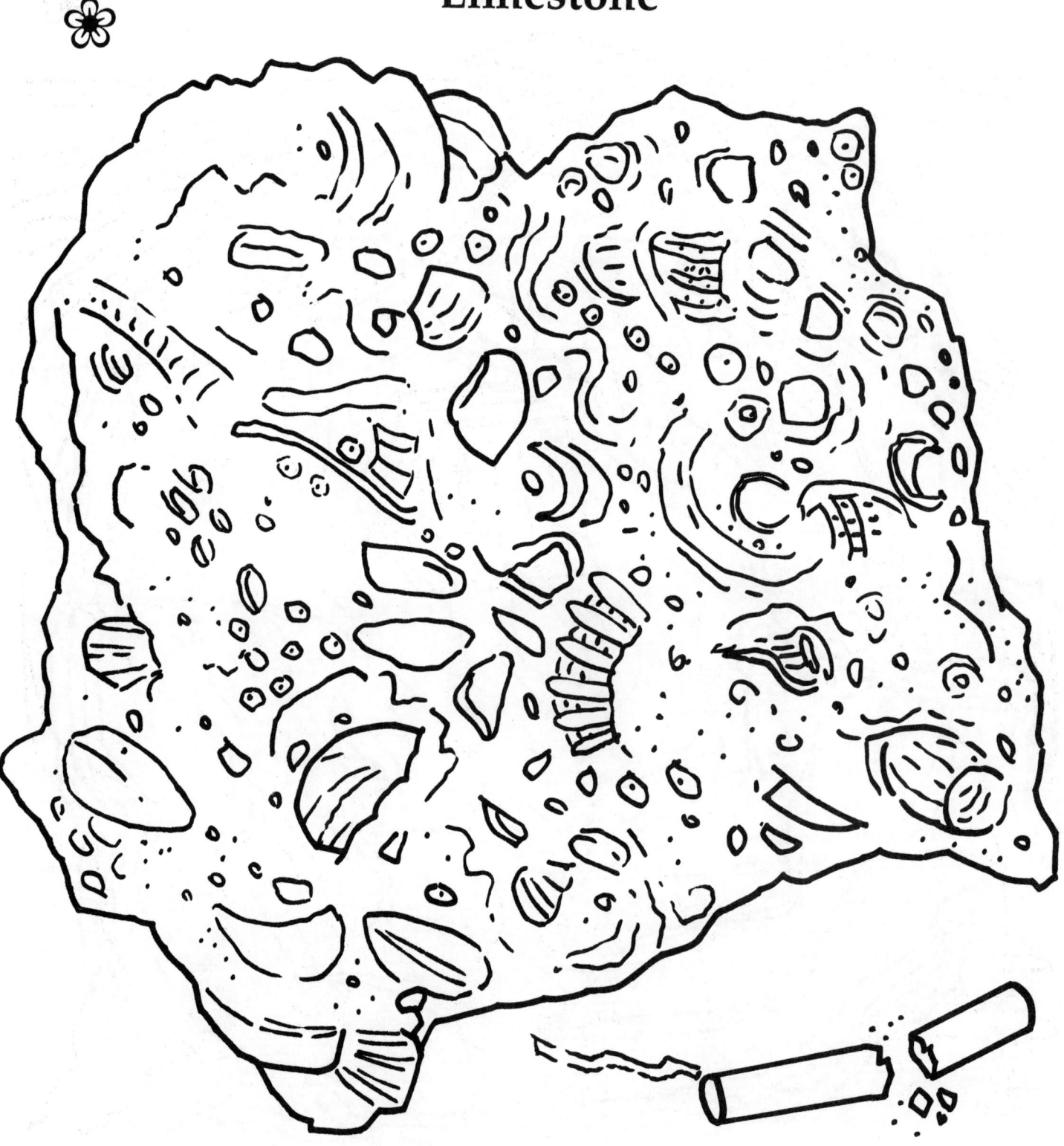

Limestone is a sedimentary rock made from the calcite left from the shells of millions of water creatures. Pieces of shell can still be seen in some pieces of limestone. Chalk is limestone that is soft and white.

Sand

Most grains of sand are parts of solid rocks that have been broken down by freezing, wind and water erosion, or the action of waves against rock. Some sands are made almost entirely of quartz or basalt grains. Beach sand usually contains pieces of broken shell.

Rock Arch

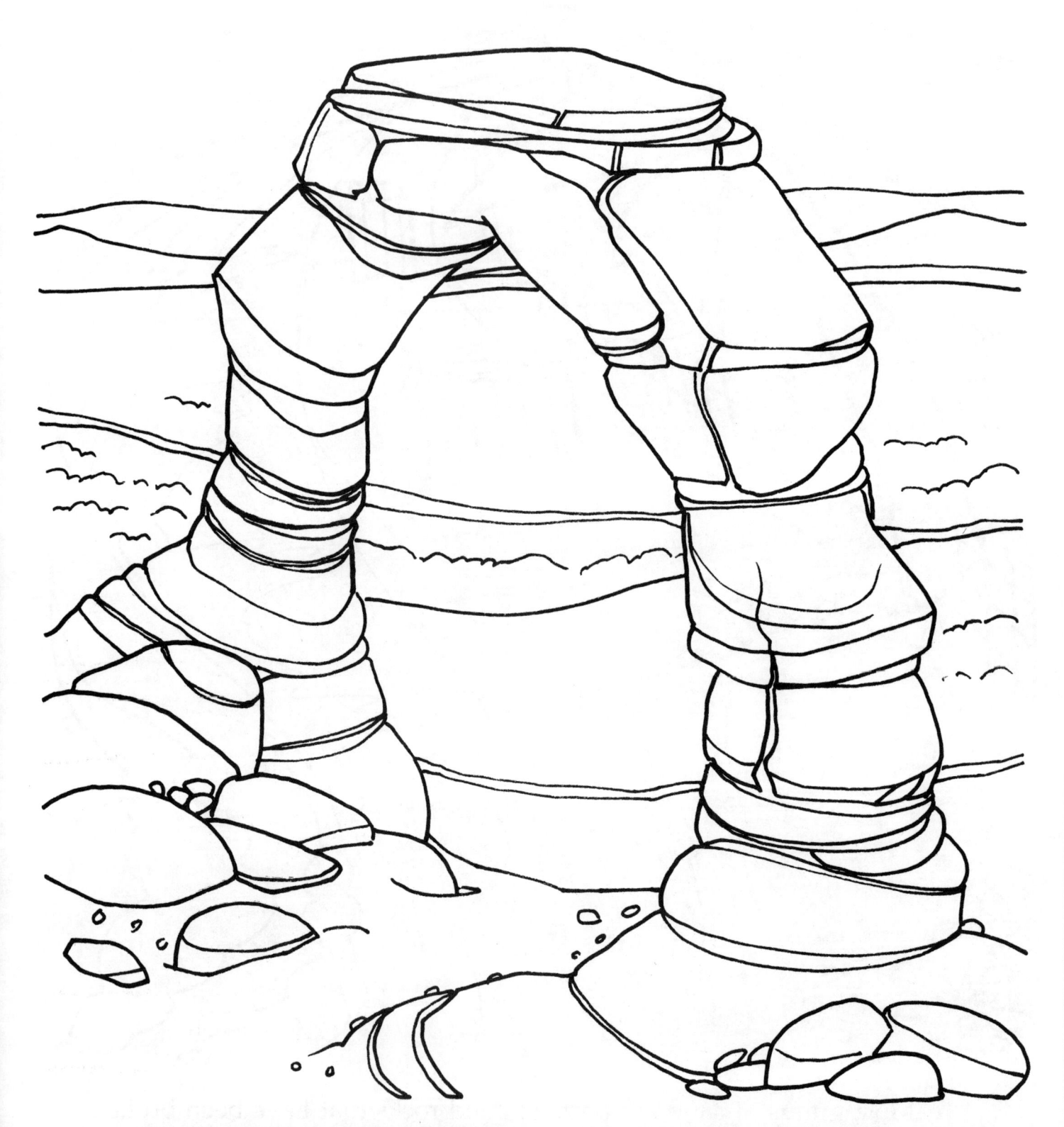

Rocks take on interesting shapes when they are affected by wind and water erosion. The movement of water or wind over a rock will wear away softer parts of the rock, leaving the harder parts behind.

Pedestal Rock

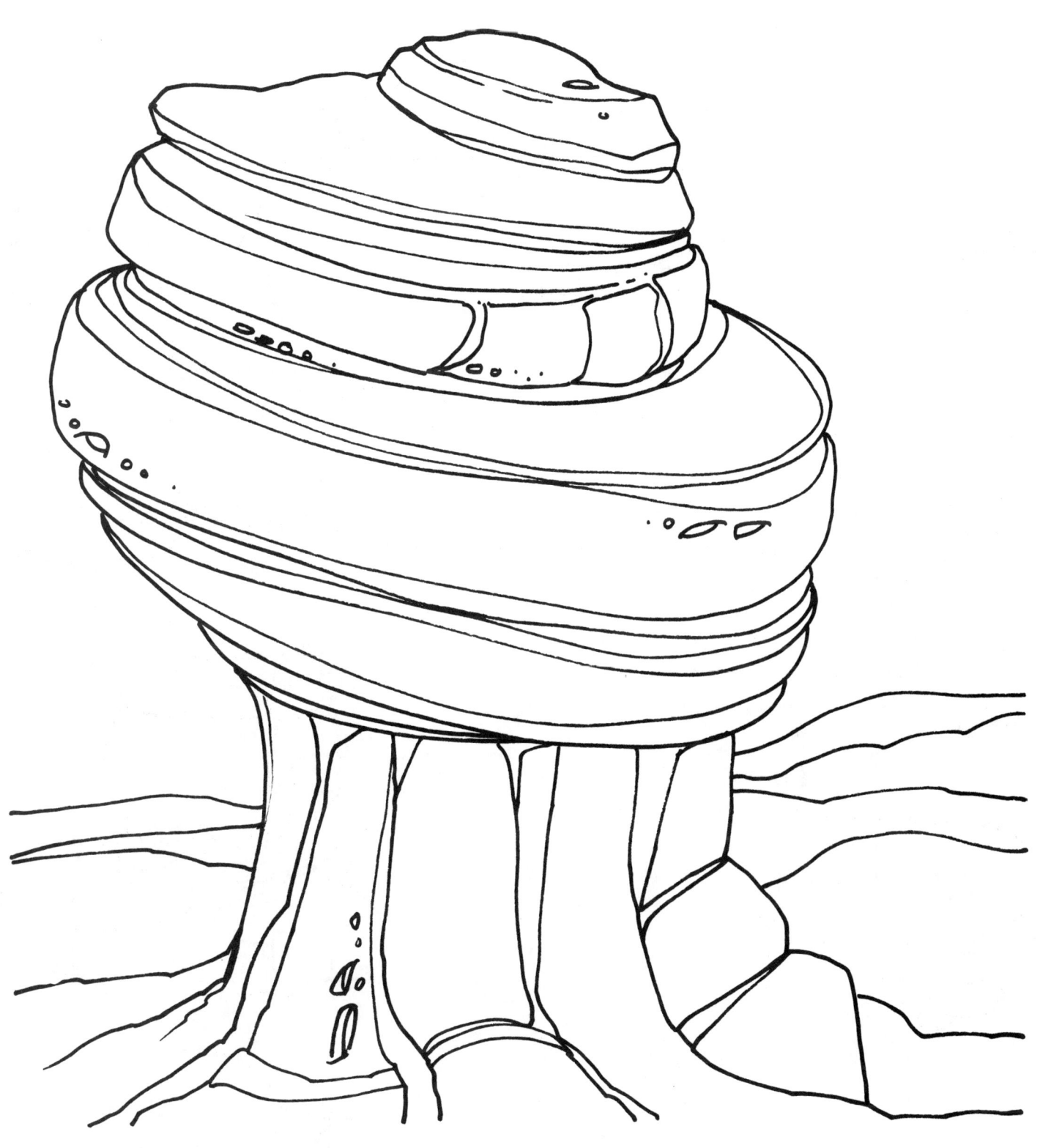

A pedestal rock is what is left when the outer soft parts of a rock are worn away by the forces of wind and water. What is left is the center column of harder stone.

Giant's Causeway

The Giant's Causeway is an unusual formation of rock on the coast of Northern Ireland. The causeway is formed of about 40,000 columns of basalt, all quite close together. Legend says that the causeway was built by giants as a bridge between Ireland and Scotland.

Ayers Rock

Ayers Rock is the visible part of a mass of sandstone that rises 1,000 feet (305 m) above the Australian outback. Although it is large enough to look like a mountain, it is one piece of red-orange rock. The distance around the rock is five miles (8 km).

Stonehenge

Stonehenge is a man-made circle of rocks in England. There are many theories about who built Stonehenge, but it has been there for centuries. Each of the 30 blocks that originally made up the circle was 30 feet (9 m) long and weighed about 28 tons (25 metric tons).

Pyramids

In the Egyptian desert travelers can still see huge structures called pyramids. The pyramids are tombs that were built by the Egyptians thousands of years ago. Each pyramid is made of massive blocks of limestone that had to be transported many miles to the building site.

Sculpture

Because they are so durable, rocks and minerals have always been used in the creation of pieces of art. Metals like gold and bronze are molded into beautiful shapes. Rocks such as granite, soapstone, and especially marble, are shaped with tools to make statues.

Mount Rushmore

Mount Rushmore is a memorial to four United States presidents in the Black Hills of South Dakota. The faces of George Washington, Thomas Jefferson, Theodore Roosevelt, and Abraham Lincoln were carved into the granite cliff with drills and dynamite.

Volcanic Rock

The lava that flows from a volcano is magma, molten rock. Lava that cools quickly hardens into a smooth, glass-like rock called obsidian. Lava that cools more slowly forms crystals, becoming rocks such as gneiss, marble, and quartzite.

Color & Learn Rocks & Minerals © Edupress EP219

River of Rock

In areas around volcanoes, geologists sometimes find fascinating rock formations that look like rivers of black rock. In these places, the flowing lava has hardened in the same shape it was flowing. It looks as if a lava flow has just stopped in one place for all time.

Aluminum

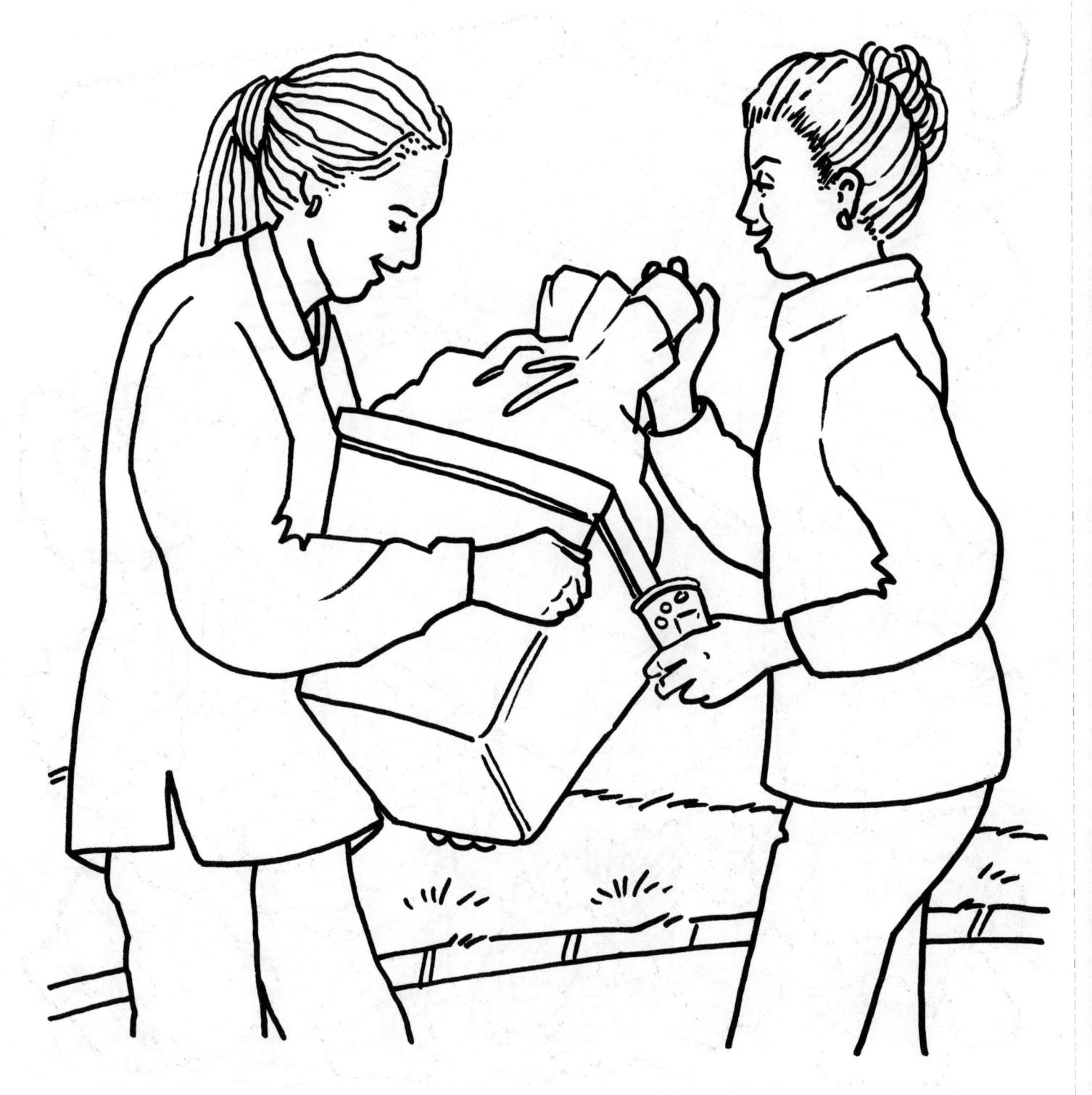

People use more aluminum than any other metal in the world. Because pure aluminum is soft, it is often mixed with another metal to form an aluminum alloy. It is more efficient to recycle aluminum than it is to mine and process new aluminum.

Gold

Gold was one of the world's first known metals. It has always been a mark of wealth. Many countries base their monetary system on gold, which is often stored as bricks, called bullion. Besides coins, gold is used for jewelry, art objects, and personal decoration.

Coal

Coal is a soft brown or black rock. It is not made of minerals, but is the remains of plant material that has been compressed and reformed under pressure. Most coal is mined from underground beds.

Petroleum

Petroleum is not really a mineral, but it is a substance that contains many minerals. It is usually found in deposits under the ground, and brought to the surface with oil wells. Petroleum products provide fuel for cars, trains, airplanes, and ships.

Water

Along with wind, water is one of the main forces that act upon rock. The movement of water causes rocks to rub against each other and against sandy river beds. This breaks the rocks down into smaller and smaller pieces, eventually forming sand.

Sorting Gemstones

Once gemstones are mined, they are sorted based on appearance, which determines their value and possible use. Less valuable samples of some gemstones, like diamonds and rubies, are used in industry. Industrial gems are not cut and polished like more valuable gems.

From Gemstone to Jewel

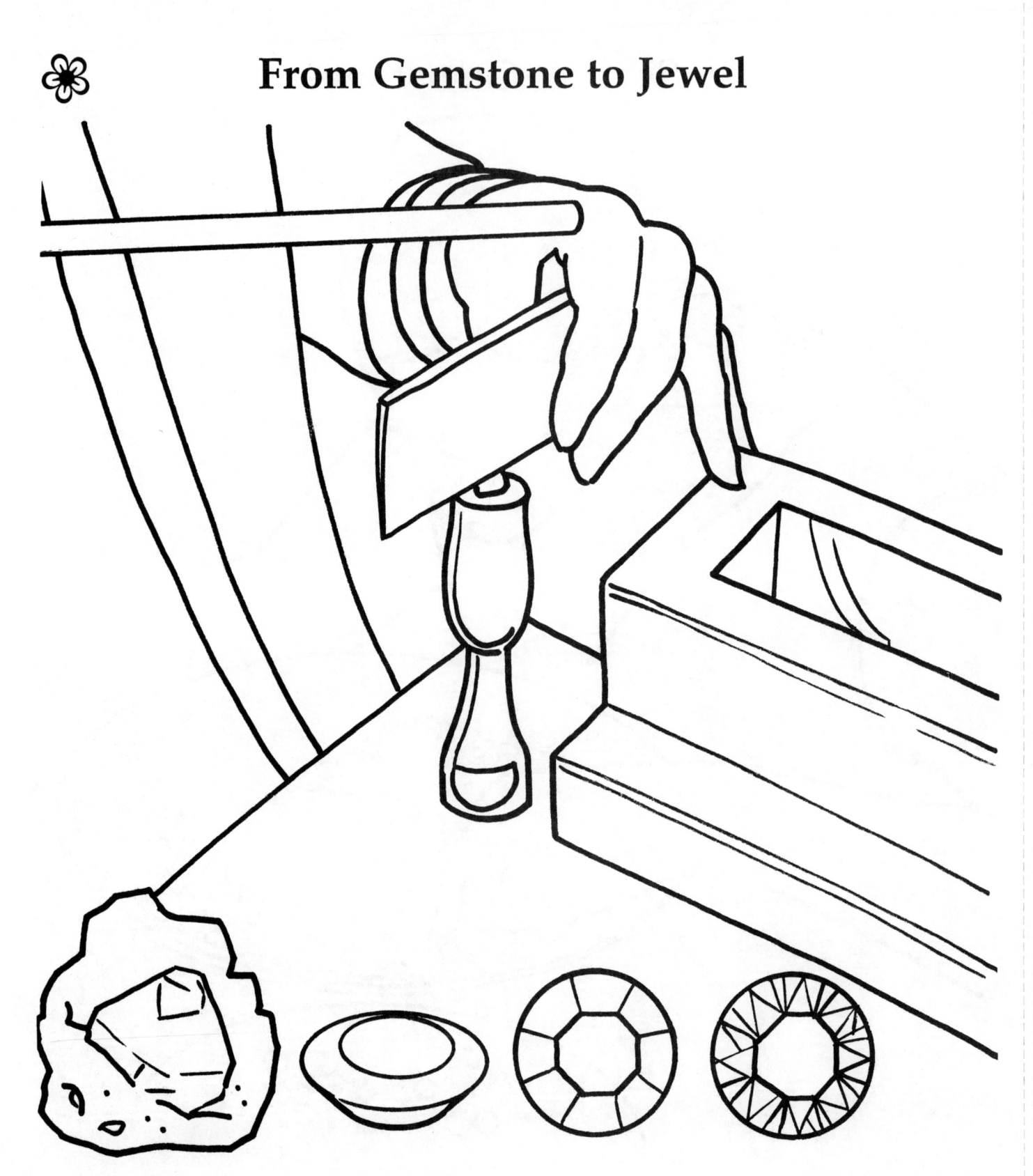

A *lapidary* uses special tools to cut a jewel from raw gemstone. First the crystals of the gemstone are ground to a round shape, with the top point cut off. Flat patches called facets are then ground at exact angles and polished. The facets reflect light, making the jewel sparkle.

Jewelry

Gems are used to create beautiful jewelry and ornaments. Jewelers combine precious metals with gems in rings, necklaces, earrings, bracelets, and other things worn by people.

Jade

Jade is a hard, highly-colored stone that is used for fine carvings and jewelry. Jade comes from two sources. Nephrite comes in a wide range of colors, including dark green, white, yellow, gray, red, and black. The more valuable jadeite occurs in the colors of light green and lilac.

Jade Mask

Jade is soft enough to be carved and shaped for jewelry, containers, and sculptures. The ancient Chinese developed the techniques for working fine jade. Pieces of jade jewelry and ornamental pieces can be found in ancient grave sites in China.

Pearls

Pearls are among the most valuable gems. While most gems are mined from the earth, pearls are formed inside the shells of oysters. When a foreign object, such as a grain of sand, gets inside the shell, the oyster covers it with layers of nacre, a shell-forming substance.